Episodes from a Colonial Present

Artists: Hangula Werner, Roshni Vyam, Michel Esselbrügge, Qi Zhou, RotmInas – Rotmi Enciso & Ina Riaskov, Maite Mentxaka Tena, Lena Ziyal

Authors: Daniel Bendix, Chandra-Milena Danielzik, Franziska Müller, Lata Narayanaswamy, Juan Telleria, Miriam friz Trzeciak, Aram Ziai

Published by Daraja Press
https://darajapress.com
Wakefield, Quebec, Canada

ISBN 9781990263460

Library and Archives Canada Cataloguing in Publication

Title: Episodes from the colonial present.
Names: Bendix, Daniel, 1980- editor.
Description: Artists: Hangula Werner, Roshni Vyam, Michel Esselbrügge, Qi Zhou, Rotmlnas – Rotmi Enciso & Ina Riaskov, Maite Mentxaka Tena. | Authors: Daniel Bendix, Chandra-Milena Danielzik, Franziska Müller, Lata Narayanaswamy, Juan Telleria, Miriam friz Trzeciak, Aram Ziai
Identifiers: Canadiana 20240284879 | ISBN 9781990263460 (softcover)
Subjects: LCSH: Postcolonialism—Comic books, strips, etc. | LCSH: Decolonization—Comic books, strips, etc. | LCSH: Equality—Comic books, strips, etc. | LCGFT: Graphic novels.
Classification: LCC JV51 .E65 2024 | DDC 325/.3—dc23

This publication is based upon work from COST Action DecolDev, CA19129, supported by COST (European Cooperation in Science and Technology)

COST (European Cooperation in Science and Technology) is a funding agency for research and innovation networks. Our Actions help connect research initiatives across Europe and enable scientists to grow their ideas by sharing them with their peers. This boosts their research, career and innovation.

www.cost.eu

Hangula
Chandra
Daniel
Roshni
Daraja Press
Friz
Ina & Rotmi
Aram
Lena
Franziska
Juan
Maite
Michel
Lata
QI ZHOU
Mute
Stop Video
Participants
14
Chat
Share Screen
Record
Reactions
Leave

Introduction and thanks

"Colonial present? Hang on, I thought colonialism had ended in... err... some time in the 20th century, right?" Well... almost, but not quite. Formal colonialism has ended, it's true: in Haiti in 1804, later in the 19th century in the rest of Latin America and the Caribbean, in India and Pakistan in 1947, in Ghana in 1957, in many other African countries in 1960, in Zimbabwe in 1980 and in Namibia in 1990. At the same time, several territories whose histories are differently related to European colonialism are still not independent, be it the French Départements et régions d'outre-mer, enclaves like Ceuta and Melilla, US outlying territories such as Puerto Rico, or the occupied West Saharan, Tibetan, Kurdish and Palestinian territories. Some settler colonial conquests such as the USA or Australia have been so successful in obscuring the whole genocidal process that they convinced most of the world (including themselves) that they are the original owners of the land.

Many people have fought and died for liberation from colonialism and their struggles were not in vain: Even the UN Declaration on the Granting of Independence to Colonial Countries and Peoples of 1960 declared that "The subjection of peoples to alien subjugation, domination and exploitation constitutes a denial of fundamental human rights" and that "All peoples have the right to self-determination" (UN 1960). The claim of racial superiority that had legitimised European rule over other peoples was, at least in principle, no longer tenable in international politics – although (former) colonizers still cling to their racialised, hierarchical prerogatives in material and ideological ways.

By the 1960s, however, Ghanaian Pan-African theorist, revolutionary and first Ghanaian Prime Minister Kwame Nkrumah (1965: 1) had coined the term "neocolonialism" to describe a situation where formally independent states have their economic systems and policy controlled by foreign actors. The structural adjustment programs associated with the IMF and World Bank loans – provided to low-income countries on the condition that they sell state assets, open up their markets and gear the economy towards paying their debts at all costs (see SAPRIN 2004) – were only the most prominent instance of "development aid" being used as an instrument of imperialism (Hayter 1971). And until today the vast sums of money extracted from formerly colonized global South countries – through debt service, profit repatriation of multinational corporations, tax evasion and illicit financial flows (the technical term for criminal elites transferring money to Swiss bank accounts) – is several hundreds of billions of dollars higher than what is being invested or given in terms of charity or "development assistance" each single year (Hickel 2017). The poor are subsidizing the rich on a massive scale.

Yet it is not only in the global economy that the effects of colonialism are still obvious: in border and migration regimes (who is allowed to travel or enjoy the entitlements of citizenship); in science (whose knowledge counts as academic, who has the funding to do research and spread knowledge); in culture (who earns royalties on intellectual property associated with music composition, literature, and arts, and whose textile patterns, beats, or stories become appropriated, whose culture counts as primitive or high culture); and even in media reporting on tragedies (whose deaths are mourned, discussed and scandalized, and whose lives are deemed worth saving). In the dominant economic, social and political settlement of our time, we can observe the continuation of deep inequalities and injustices established during 500 years of European colonialism. The Peruvian sociologist Aníbal Quijano (2000) talks about the "coloniality of power", meaning that its distribution within capitalism is shaped by the race, gender and class divisions of colonialism. US-American Black feminist theorists like bell hooks (1981) and Patricia Hill Collins (1990) have highlighted the need for feminist liberation to avoid a reductive, singular category of 'woman' that is meaningless without an understanding of colonial race and class divisions. Indian postcolonial theorists like Gayatri Chakravorty Spivak (1988) and Chandra Talpade Mohanty (1988) provide reminders that the way the world is seen is through the lenses of 'our' European oppressors, and to know ourselves is to take our own positionalities, our own histories as starting points rather than the dominant histories written for us.

It is from these struggles that we take our inspirations for the stories within. And even as we brought these stories together from a place of dialogue and hope, we were contending with our own crises and associated struggles early on in our collaboration. We came together to establish this collaboration at the height of the COVID-19 pandemic in mid-2020, the majority of us in lockdown in our towns and cities all over the world, and with COVID-19 rapidly spreading around the world, confining us to our homes and overwhelming our health systems. Amidst the devastation we observed that yet again the coloniality that is so much a part of our everyday was a determining factor in whose knowledges counted, who was vaccinated and who was not, who lived and who died.

Our collaboration also offered hope: through art, through dialogue and through solidarity. Many of us have never met in person and yet we feel a deep sense of commitment and loyalty both to each other and to ensuring that we do justice to the stories we're telling. In our collective fight against the colonial present we recognize that new technologies offer possibilities to build solidarities and pluriverses across geographies and temporalities. We are struck by both a sense of helplessness but also hope that our collaboration has brought us – that working through the pandemic has been both dispiriting but also generative, that in our partnership with each other and with the artists we had convivial debates about our own positionalities, but also sadness that we are unable to do justice to all of the stories that need to be told about the histories of colonialism and their contemporary effects all over the world.

And then in the midst of finalising this comic, even more stories have emerged to which we are unlikely to be able to do justice. As of July 2024, we have been witnessing for almost nine months Israel's assault on Gaza, an escalation of violence triggered by the killing, maiming, rape and abduction of more than a thousand people in Israel by Hamas. Israel's ethnic cleansing and plausible genocide of Palestinians in Gaza and the intensification of settler-colonial violence in the West Bank have been accompanied by a surge in anti-Palestinian racism, anti-Muslim racism and antisemitism across the globe. With the attention on Gaza, the violence and displacement taking place in other places farther away from us in Europe, including Sudan, DR Congo, Haiti and too many other places besides, remain largely hidden from Western journalistic view. That we watch in horror as these armed conflicts – each with their own colonial connections – play out in real time in ways that are violently killing many people, even as governments seek to suppress our right to know about, or protest against, such atrocities, is too often unbearable. Where do we focus our energies and which stories do we tell? So here we are focused on how our own activist lives interact with our positionalities, telling stories that we hope opens the doors to many more people being able to tell their stories.

In this context, we believe that our stories, entangled as they are in the more recent European imperial past, stress the importance of avoiding a rivalry of victims or a hierarchy of crimes between, for example Shoah versus Maafa, two crimes against humanity with which our own histories are intertwined and which we bring into dialogue in one of the stories in this book. It is crucial to explore the entanglements between genocides and the productive "multidirectional memory" (Rothberg 2009) linking them, and to plead for joint efforts to investigate and condemn hatred against, for instance, Jewish or Black people as much as other racisms. We decided to do this book to draw attention to postcolonial inequalities and struggles for justice in a world marked by colonial legacies; to capture colonial echoes even in strange places; and to reflect on our own individual connections to colonialism. However, instead of writing another, dull and hard-to-access book with tons of text (and thus again recreating academic hierarchies), we set out to sail in a different direction and sought to develop this illustrated comic book. Initially, we wanted to do a comic on postcolonial theory in the narrow, academic sense. But who wants to read a nerdy drawn conversation between prominent post-colonial authors like Edward Said, Homi K. Bhabha and Gayatri Chakravorty Spivak around a dinner table? (Apart from us, that is.) So, we decided to make it more personal, to see how we are connected to colonial legacies, in our families, everyday lives, as well as our academic and activist struggles. We sought to challenge how we know, expanding our imaginations through visual stories, and telling these stories from what Kimberlé Crenshaw (1989), known for establishing the concept of intersectionality, would argue is our diverse, intersectional identities, where we contend variously with our own past and present experiences of discrimination, privilege and struggle, whilst contemplating the possibilities these stories hold for forging more just futures.

All of us were involved as editors in developing each story. Most artists have a particular connection to the topic of the story or the particular context in which it is set. While the authors came up with the stories, the long process of developing the stories meant that the lines between authors and artists sometimes became blurred. Some authors became part of the creative process and some artists participated in the making of the stories.

"Never Conquered. On the Destruction and Creation of Knowledge" is inspired by friz', Ina's and Rotmi's experiences of queer_disidencia sexual_lesbian affinities and the transnational solidarity movement with the Zapatistas and other indigenous and social movements between Mexico and Germany. It revolves around the question of how marginalized knowledge can be made visible, as well as how, within the context of colonial and heteronormative power dynamics, other forms of solidarity and care can be formed to create and initiate justice.

"It All Runs in the Family" is based on Franziska's research into her family history, especially the biography of Louis Hanitch, lawyer and Deputy Attorney General in Dakota Territory (in what is now the US). She realized that her ancestors not only contributed to National-Socialist (Nazi) atrocities, but also contributed to the commodification of Dakota land by law enforcement and introducing new borders. This process made her realize how German settler colonialism is romanticized, and how a "white Atlantic" separates and whitewashes the stories of German emigration and settler colonialism. As part of the transnational struggle against fossil extractivism, she experienced encounters which unite yet also separate the North Dakotan and Hambach Forest activists.

"Tracking Trauma. German Genocides at Home and Abroad" is inspired by research into Daniel's family history. Members of his Jewish-German family took part in German colonialism and the genocide against Ovaherero, Nama, San and Damara peoples in what is now Namibia. While many perished in the Shoah, others managed to flee or migrated to South Africa after surviving National Socialism (Nazism). Today, his family owns a farm in Namibia. Hangula's background also influenced the story's perspective on the German genocides and their legacy. He had to flee Namibia as a child during South African colonisation and lived as a refugee in the German Democratic Republic. Upon his forced return to Namibia as a teenager, he lived with a German family and eventually went to the German School in Windhoek. Particularly in the context of the quest for justice for colonial genocide in Namibia, he became acutely aware of the problematic contemporary postcolonial power relations between Namibian communities of Colour.

The third intermezzo "Whose Cup of Tea. Migration, Colonialism & Plantation Capitalism" is inspired by Chandra's Sinhalese family background. Over the last half a century, many family members have migrated to Germany and Saudi Arabia to ensure their families' survival or to be better off. There, they have been subjected to physical, emotional and exoticised sexual exploitation in the low-wage sector, in care work, and due to restrictive migration and oppressive deportation policies (here, there is also a connection to the first intermezzo "Savages Setting Sail: Gender, Sexual Violence & Colonialism"). At the same time, she has grown up in a German society, which indulges in "Ceylon" tea (cultivated by her petty farmer family members as cash crop for export) and praises the marvelous effects of Ayurvedic massages and Yoga retreats in the cheap holiday destination of Sri Lanka. Her aunt died as a child due to hunger inflicted on Sri Lankan society by colonial and (neo)liberal subordination to global capitalism.

"Alienating the SDGs. A Critique from Outside" is inspired by Juan's experience working with different

NGOs, both in Europe and in Latin America. Two aspects caught Juan's attention during this period. First, on many occasions, those who actively promote the current unfair economic and political global system and those who suffered the severe negative consequences of this system tended to use the same 'buzzwords' (see Cornwall and Brock, 2006), including 'development', 'participation', 'freedom', 'capabilities' and 'resilience', amongst many others. Second, this discourse was generally put forward by an institution that most people tend to accept uncritically: the United Nations. These aspects triggered Juan's curiosity about the way the United Nations constructs its discourse about global issues.

The last intermezzo, "Blurred Identities. Can I be the Coloniser and the Colonised?", captures a small yet significant encounter that Lata experienced as part of a 'voluntourism' experience in Central America when in her 20s. Being both Canadian and visibly of a darker-skinned South Asian ethnic origin clash in ways that unsettle 'civilisational' divides marking the way we understand notions of 'development'. It is a reminder that people in racialised bodies are never free, neutral nor independent, raising questions about how minoritized bodies 'belong' in the 'West'. Their lives are always filtered through the lens of colonial racialised hierarchies that can never be dislodged nor forgotten, not even in their private leisure time.

"Under Development. Future Uncertain" is inspired by Aram's, Daniel's and Lata's diverse experiences: campaigning and demonstrating against the Sardar Sarovar dam in the Narmada valley, participating in the anti-globalisation protests and in People's Global Action; involvement in the degrowth movement; work in the development industry; and teaching post-development. The artist Roshni's artistic expression is grounded in her experience of the suppression of tribal people and their art forms in India.

The stories often deal with violence, sometimes in extreme forms such as genocide. This is of course no surprise as colonialism as such is a crime against humanity, albeit not officially recognised as such. Readers will, however, be touched differently by the stories, depending on their own position within the legacies of colonialism and its accompanying mass violence.

All of the contributors to this book (editors/authors and artists) have bundled together our acknowledgements on the basis that without our varied connections, we would not be who we are and in turn would not have collaborated on this project in the way that we have. So bringing them together allows us to share in the inspirations of our fellow contributors creating a collective spirit in which all are involved and benefiting from our cooperation and mutual engagement in the process of bringing together this collection of graphic stories. Thanks to the following for inspiration, feedback, money, admin work etc.: Daraja Press and especially Firoze Manji for jumping into cold water with us (it is the first comic book by Daraja Press) and Kate McDonnell, Parastu Karimi, the Auenland hobbits of Kassel, Durga Bai, Mary Ixmal Bautista, Mark Bendix, Paul Bendix, Andreas Bohne, BUKO, Juan Caamal, Cape Town Holocaust & Genocide Centre, Congreso Nacional Indígena, Helene Decke-Cornill, Brigitte Danielzik, Edith Danielzik, Heinz Danielzik, Degrowth Movement, Katharina Döbler, Teboho Edkins,

Ende Gelände, Arturo Escobar, Global Tapestry of Alternatives, Friederike Habermann, kassel postkolonial, Luz Kerkeling, Reinhart Kößler, Janne Lahti, Urs Lindner, Hans-Christian Mahnke, Mauricio, Linda Mbeki, Henning Melber, Christel Müller, Jörg Konrad Müller and Lydia Müller for keeping original documents that allowed tracing the Müller/Hanitch family, Narmada Bachao Andolan, Christian Obermüller, Peoples' Global Action, Laura Cristina Revilla Sanchez, Wolfgang Sachs, Jana Schäfer, Erik Schnack, Helena Scully Gargallo, Thomas Simoes, Late Jangarh Singh Shyam, Christoph Spehr, Ángel Sulub, Unsettling America collective, Subhash Vyam, the Zapatistas, Jürgen Zimmerer, all people who made possible La Gira Zapatista in 2021, all colleagues in both the Convivial Thinking and Decolonising Development collectives, with special mention to Julia Schöneberg for being such a kindred, decolonial spirit, our families, our friends who are family, and all the named and unnamed people who inspired these stories in their struggle for a dignified life for all in the past and today.

NEVER CONQUERED

ON THE DESTRUCTION AND CREATION OF KNOWLEDGE

AUTHORS: MIRIAM FRIZ TRZECIAK & ROTMINAS
ARTISTS: ROTMINAS - ROTMI ENCISO & INA RIASKOV

In 1562, Fray Diego de Landa, a Spanish Franciscan, had all Mayan writings and images burned. Very few of the documents survived. Years later de Landa became bishop of Yucatan, Mexico.
As an inquisitor, he relentlessly pursued, tortured, and executed many people who refused to convert to the Catholic faith.

Templo Mayor, Tenochtitlán.

500 years later...
HOSTAL
BOXITO
Merida, Yucatan, Mexico, Boxito Hostel.

HOSTAL BOXITO.
HOME FAR FROM HOME
THE CASTE WAR OF YUCATAN
LOS MAYAS CRUZOOB
Alejandro always greets me, and we've become friends since I arrived here. During his holidays he works here at the Boxito Hostel. He is studying social anthropology.
Johanna is from Germany. She came to do research for her thesis. I see her immersed in her books all the time.

Wow, I thought you were on vacation.
The Caste War (1847-1935) was one of the largest conflicts between the rebellious Mayan population and the local white and mestizo elites on the peninsula. However, some mestizos also fought on the side of the resistance.
THE CASTE WAR
LOS MAYAS CRUZOOB
LA GUERRA DE CASTAS 1847-1935

I am researching the history of indigenous resistance here in the region.
I see, the Caste War.
Yes, but I am mainly interested in gender relations. ... Oh, I need to run, I have an appointment at the university to conduct an interview with an expert on the topic.
Hold on a sec ... if that's what you are interested in I could tell you something.
Two of my ancestors were Cruzo'ob when María Uicab ruled Tulum ...

My grandmother often told me about our Cruzo'ob ancestors and the patron saint, a tough and wise woman.
María Uicab was the political, military and spiritual authority of the rebellious Maya of Tulum between 1863 and 1875. She interpreted the talking cross, a sacred symbol, for her people. Under her rule, Tulum became the centre of the Cruzo'ob struggle. Cruzo'ob is Mayan and means "crosses". It is also the name used to refer to the Mayan insurgence during the Caste War.
To this day, people on the peninsula continue to practice the cult of the Talking Cross.

In my village there is still the tree where the insurgents were hanged during the Caste War.

[1] First Declaration of the Lacandon Jungle
https://radiozapatista.org/?p=20280&lang=en

On 12 October 2020, a group of organized Otomi indigenous people in Mexico City took over the facilities of the National Institute of Indigenous Peoples (INPI) to demand housing, health, work and education.
http://rip.mx/one-hundred-days-of-otomi-hope/
"democracy, freedom and justice"

Kenia Inés Hernández Montalván from Guerrero, Mexico, is a lawyer and Amuzga indigenous activist. She has been dedicated to the legal and political defense of women victims of violence, as well as to the defense of territory and the rights of indigenous peoples. Since 2020, she has been a political prisoner. Her life in jail is in danger due to her delicate health condition.
https://www.frontlinedefenders.org/en/profile/kenia-ines-hernandez-montalvan
Tw: @ParaKenia
"Feminicidal Mexico, Sister Kenia here is your pack, #FreeKenia"

Cherán a purepecha indigenous town in Michoacan, México suffered from organized crime, corrupt politicians and police. Kidnappings, extortion, murders, and illegal logging of the local forest were part of daily life. In 2011, the people of Cherán started a revolt led by women and installed self-governace that led to peace and justice.
https://en.wikipedia.org/wiki/Cherán
"Cherán Keri in defense of our woods"

Berta Cáceres Flores was a Honduran Lenca indigenous environmental, human and women rights activist. She was the general co-ordinator of COPINH (Civic Council of Popular Indigenous Organisations). She was assassinated on 2 March 2016 in her home by armed intruders, after years of threats against her life.
more info: https://copinh.org/category/english-es/
"Justice for Berta Caceres!"

In El Estor, Izabal, Guatemala, Q'eqchi' communities continue to resist a destructive nickel mine despite growing state militarization, killings, and arrests.
https://www.facebook.com/EstorResiste
"We will defend our territory. El Estor resists #TheMineKills"

Samir Flores Soberanes was a nahua indigenous journalist, environmental, land and territory human rights defender. He was shot dead on 20 February 2019, after receiving several death threats for his work defending the territory of his community.
more info: https://www.samirvive.art/home
"#SamirIsAlive"

Johanna is so tired, she falls asleep. In her dream she meets Diego de Landa.
You are not allowed to rewrite history.
WITCH! I'm going to have you burned at the stake!
I'll write the paper and you can be my co-author.

Do you know that one of the four manuscripts that survived the epistemicide is in Dresden, Germany?
Really? What a coincidence, I can go there and send you a picture.
Ugh, what a nightmare!
A photo, haha, no thanks. We don't want photos or charity, we want JUSTICE!!!

Lufthansa
LH 497 19:05
FRANKFURT
A5
Johanna returns to Germany and Alejandro goes back to studying.
Lufthansa

[2] https://enlacezapatista.ezln.org.mx/2021/04/20/421st-squadron/

Banner 1: "DB (national railway company of Germany) out of Mexico! No to the ~~Mayan~~ colonial train!"

Banner 2: "National Indigenous Congress It is the time of dignity."

In the meantime,
at the Saxon State and
University Library Dresden (SLUB).

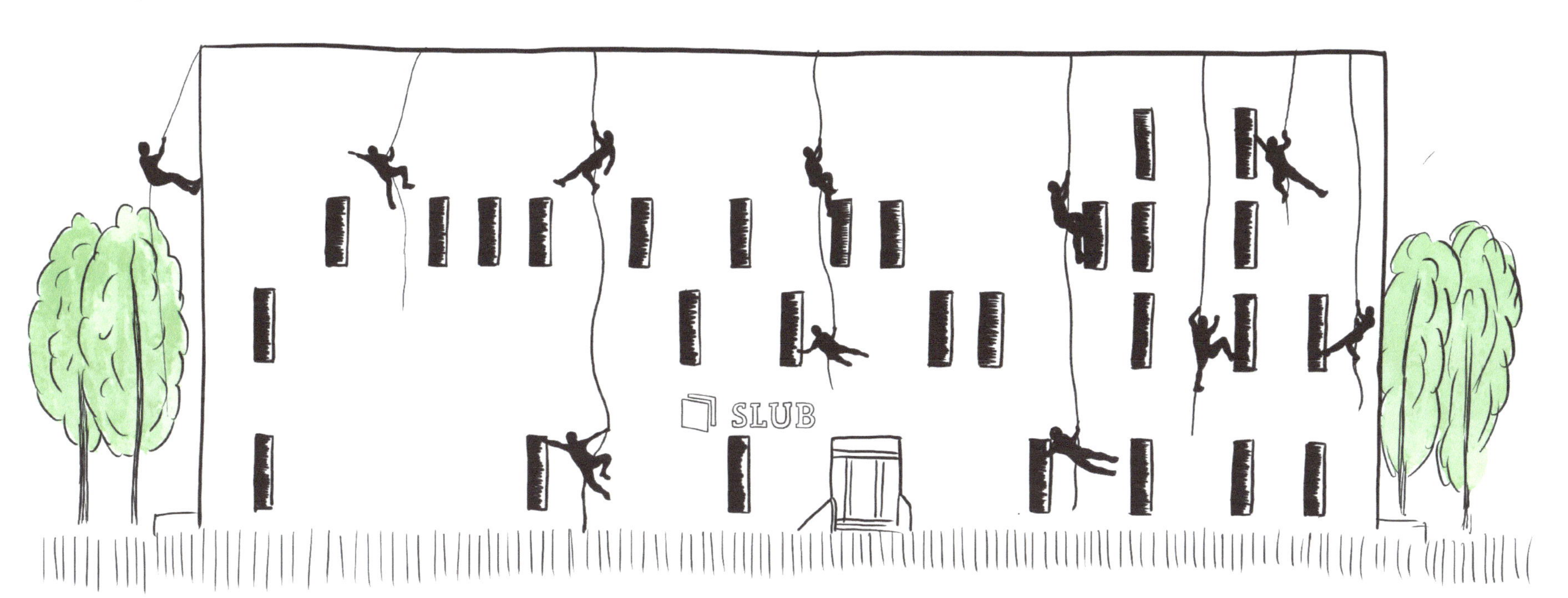
SLUB

To be continued...

Savages Setting Sail

Gender, Sexual Violence & Colonialism

Author: Chandra-Milena Danielzik
Artist: Michel Esselbrügge

EUROPEAN COLONIALISM WAS NOT ONLY DRIVEN BY MILITARY OCCUPATION AND ECONOMIC EXPLOITATION, BUT ALSO BY SUPREMACIST IDEOLOGY, HYPOCRITICAL MORALS AND FANTASIES ... PATRIARCHAL AND SEXUALISED FANTASIES.

THE LATTER WERE KEY TO LEGITIMISING SAVAGE CRUELTIES - VIOLENCE, RAPE, MURDER, ENSLAVEMENT ... AND THE DESTRUCTION OF WHOLE SOCIAL AND ENVIRONMENTAL SYSTEMS.

I'M COLUMBUS THE FAMOUS EXPLORER AND I KNOW THAT THE WORLD ISN'T ROUND! IT'S RATHER SHAPED LIKE A WOMAN'S BREAST. THE NIPPLE IS ITS HIGHEST POINT, CLOSEST TO THE SKY. THAT'S THE DESTINATION OF MY JOURNEY!

HAHA! WHAT AN IDIOT!

IT MIGHT SEEM AMUSING AT FIRST, BUT THIS IDEA OF A JOURNEY ON A WOMAN'S BREAST IS PART OF WHITE MEN'S COLONIAL FANTASIES. IT CANNOT BE SEPARATED FROM SEXUALISED VIOLENCE AGAINST RACIALIZED PEOPLE.

COLUMBUS AND HIS MEN RAPED TAINO WOMEN AND CHILDREN, AND ENSLAVED AND SEX TRAFFICKED THEM FOR FURTHER EXPLOITATION. PERFIDIOUSLY, THEY FANTASISED ABOUT FEMALES AS SEXUAL SINNERS AND HENCE ABUSABLE.

BY THE WAY, UNTIL THE MID-TWENTIETH CENTURY MOST EUROPEAN STATES ONLY CRIMINALISED THE RAPE OF WHITE WOMEN.

THE SYSTEMATIC RAPE AND MASS MURDER OF WOMEN WAS INTENDED TO DESTROY THE LIFE OF INDIGENOUS NATIONS. THIS GENOCIDE INCLUDED UNTHINKABLE ATROCITIES: PEOPLE - EVEN INFANTS - WERE FED TO DOGS.

BUT HOW WAS IT POSSIBLE FOR THE EUROPEAN COLONISERS TO HAVE SEXUAL FANTASIES ABOUT INDIGENOUS PEOPLE AND AT THE SAME TIME DESPISE AND BUTCHER THEM?

...THESE SEEMING OPPOSITES ARE PART OF THE SAME PICTURE, AS YOU CAN ALSO SEE IN A EUROPEAN ENGRAVING OF VESPUCCI "DISCOVERING" THE SUPPOSEDLY VIRGIN LAND: AMERICA.

Longing for the breast
infant (can't survive without)
man (sexual conqueror)
desire
The Americas, Asia, Africa, Caribbean, Australia (nurturing virgin territory; exoticised & sexualised)
Just like woman, land must be possessed and dominated by white men.
The "discovered" land was considered "Terra Nullius" by the Europeans.

IT MEANS "NOBODY'S LAND". THIS LEGAL EUROPEAN CONCEPT ALLOWED COLONISERS TO CLAIM LAND EVEN IF IT WAS INHABITED. IT WAS DECLARED EMPTY 'CAUSE ITS INHABITANTS (A) WEREN'T CONSIDERED PART OF HUMANITY BUT "WILD" AND PART OF NATURE AND (B) HADN'T "PRIVATIZED" IT.

America, engraving by Jan van der Straet, 1580

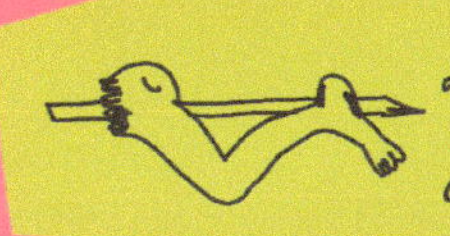

indigenous people depicted as cannibals, savage, wild, dangerous...

Strange "exotic" animals

Cross: Christianity, faith

JUST LIKE PEOPLE AND TERRITORIES, WE WERE CLASSIFIED, SUBORDINATED, EXPLOITED, COMMODIFIED AND EVEN WIPED OUT.

IN THIS EUROPEAN PICTURE OF A SO-CALLED DISCOVERY, I'M SUPPOSED TO REPRESENT AMERICA. OH MY GODDESSES! WHY AM I LYING IN BED NAKED AND APPEALING, WHILE MY FRIENDS IN THE BACKGROUND ARE DEPICTED AS WILD SAVAGES? TAKE A CLOSER LOOK AT THIS EXQUISITE EXAMPLE OF HOW EUROPEANS CONSTRUCTED THEMSELVES AS VIRTUOUS, CIVILIZED AND PROGRESSIVE— HOW THEY FANTASISED ABOUT THOSE THEY OPPRESSED AS THEIR ABSOLUTE OPPOSITES - LUSTROUS AND OVERSEXUALISED, BARBARIC AND BACKWARD. THIS IS HOW THEY LEGITIMISED MURDER, RAPE AND PILLAGE.

WHAT A CREEP!

MANY INDIGENOUS SOCIETIES HAD DIFFERENT CONCEPTS OF GENDER, FAMILY AND SEXUALITY THAT WENT BEYOND THE WESTERN PATRIARCHAL, HETEROSEXIST IDEOLOGY OF ONLY TWO BIOLOGICAL SEXES. SOME WERE GENDER-FLUID, SEX-POSITIVE, NON-HETEROSEXUAL, NON-HETERONORMATIVE AND HAD TOTALLY DIFFERENT FAMILY STRUCTURES. THESE WERE DESTROYED THROUGH GENOCIDE, CULTURAL AND RELIGIOUS DOMINATION, PHYSICAL AND MENTAL VIOLENCE AND EUROPEAN LEGAL SYSTEMS.

armour

armed

Vespucci not naked, standing → erect
male = rational
Astrolabe (astronomical instrument) CULTURE
= Knowlege → European rationalism

VS.

Indigenous Woman
naked, receiving, passive
→ alluring, sexual
NATURE
female = emotional irrational

Encounter = heterosexual, binary

Heteronormativity is a European/Western fantasy!

IT ALL RUNS IN THE FAMILY

Author: Franziska Müller
Artist: Qi Zhou

Hamburg, 1980

Let's see...
There! That's him in the picture. Ah yes, great uncle Charlie...

His father was a lawyer and he always made sure we got generous amounts of chocolate and chewing gum.
You know, we had nothing in those days after the war.
What does a lawyer do, Opa?
Hm... to put it simply...

He makes clear what is just and fair.
Wow! Can we visit the lawyer?
Well, I'm afraid he doesn't exist anymore. Neither does great uncle Charlie.

Then, can we go to America?
Oh, Sonja. I'm sorry, but we can't.

I mean, we would have to fly, and that's just sooo expensive.
Aw...
But we could go hiking in the Alps instead!

But I wanna go to America!
I wanna know what the lawyer did and why they had so much chocolate!
'Cuz I want a lot of chocolate, too!
There, there. If you want some chocolate, Sonja...

...I have some for you, right here.
Want some?
Yaaaay!!

This is...

...Where her story began.

Protest camp in the occupied Hambacher Forest, 2018

Hey, you guys wanna see it, too?
Yeah, sure! Thanks.
...
CRACK

Sonja, wanna look, too?
Oh yes, what is it?

It's a donation from Indigenous activists from the Dakota Access Pipeline in the U.S..
They say they've woven their prayers into it, to 'provide us with all the strength and endurance we need for the fight'.
Ooh.

That's really nice of them. We are all one!
I guess so, I don't think it actually works though...
That's not what it's about!

But, speaking of Dakota... some relatives of mine actually lived there.
A great-great-grand-uncle, I think.
Huh, the world sure is small.

Isn't it? Maybe the struggles really are connected.
It's a bit corny if you say it like that, haha.
Hey, don't mock me!
STANDING R
Are we all one? All these liberal ideas...
TAND WITH STANDING ROCK
You know, I'm really intrigued now.
I always wondered what my relatives were doing there. Maybe I should try to find out!
Solidarity, that's much more than slogans.
Hey, who took my chocolate?
Museum of Emigration, Bremen, Germany, 2018

...and up until 1890 the number of German emigrants living in the US rose to 2.8 million.
And that's because...?

Well, they were facing such miserable conditions.
Farmers were very poor, some were religiously discriminated against.

You see...
They were all hoping for a better life in the land of the free!

Hm, really...

So they did just the same in the United States? Just farming?
Well, many of them were farmers, others worked for the railway, some followed the gold rush.
I see...
We're at the end of our tour. I hope it could provide you some insight.

Thank you. I learned a lot.
I think I'll try to find out even more.

Museum of the American Indian,
Washington D.C., 2019

I've learned that many German people immigrated to Dakota.
I assume they did similar things as they did in Germany, like farming?
Well, that's one way to put it...
Well, they thought they could just settle on the land of our Dakota Nation.

The European settlers tried to remove us by breakingthe treaties and clearing the land for them.
They imposed this odd idea that land can be bought and sold and belong to individual people.

Oh, wow... I didn't know.
I never heard it like this.

It's good that you do now. Here you can see how settler privilege works.
There's a lot to learn and it will be uncomfortable, but you need to take this first step.

You're right. Thank you.

Anishinaabe

Now, there's another part of this settler-colonial story that got lost in the Atlantic...

Bismarck, North Dakota, 1889

...Ah, there it is.
'Louis Hanitch now appointed to Assistant Attorney General for the Territory of Dakota'.
The Bismarck Tribune.
Couldn't have written it better myself!

KNOCK KNOCK

Seems my guests are punctual to the minute.

Look who it is! It's been a while.
How are you, Mr. Attorney General of Dakota?
Congrats, dear Louis!

Thank you boys. It's 'Assistant' Attorney General, by the way.
Hah!
Isn't it all the same? We know there's no stopping you! Just like in college...

Anyway, we brought something to celebrate...
Huh!
A special guest, directly from a Florida swamp!

He's all yours, Louis!
Ugh... Why must I be dragged into this...
An alligator?! What an insane idea! I love it!

Well hello, little fellow. Applying to be my Law Enforcement Officer, eh?

No. Back off!
Woah!

Are you alright, Louis? I'm so sorry, I'll get rid of him!
Yikes.
Don't sweat it. I like the fellow's spirit! He's a keeper.

Hah! I'm just blabbering. Let's bring some order to Dakota, boys!
Our new colleague... 'Bismarck' can watch the house while I'm away. Haha!

But if you ask me, my friends... There's no better way to celebrate than to keep up the good work!
I say we go keep the unruly Indians at bay over here at the Northern Pacific Railway.
Have a wonderful day, Mr. Hanitch.
Thank you. Take care of the alligator.
Geez, Louis, are all Germans like this?

You know, old pal, sometimes it's not so important where you're from...
...But where you are.
?

North Dakota, 1891

What's going on here?
It's that German lawyer Hanitch giving a speech on our economic prospects.
He's got an alligator!

Mr. Hanitch, now that we've finally gained control over the land of the Oceti Sakowin people...
...What's your view on the new railroad and on our economic challenges?

The wheat question is the foremost just now. We need to exploit these fertile soils to excel as a wheat exporter.
A week ago last Monday 1,600 cars of wheat were received, the next day 1,200 cars, and it is averaging about 1,200 a day. That is a lot of wheat.

North Dakota wheat is the best, and we want it,
and it is only a question of time when we will maximize our yields.
CLAP
CLAP
CLAP
Ambitious, isn't he? Just keep growing wheat on stolen land. Eventually the soil will fail you.

Superior, Wisconsin, 1919
Today, we hope to finally resolve the pending case of the water boundary between Minnesota and Wisconsin.

We welcome Mr. Louis Hanitch, speaking on behalf of the state of Wisconsin.

Thank you, your honor.

Borders are the essence of a Nation. Dividing Lake Superior amicably between Minnesota and Wisconsin will serve the best interests of our federal states.
I present you the following proposal: drawing the border across the waters of Lake Superior.
MINNESOTA
Lake Superior
WISCONSIN

The best interest of Minnesota would be to expand our control over Lake Superior.
Hmph...

Everybody would argue like that. But don't we share a common interest: control the lake?
Let's do this in a mutually benefical way, and keep these Indians away from our harbors and shores.

The audacity! Arguing over land that is not theirs.

As if it's the most natural thing to drive people away from their homes, who have always lived in harmony with the land!

The following decree is issued: The water boundary is drawn in accordance with the plaintiff's map.
May our sister states thrive! And may Lake Superior be forever remembered by this name.

Right. Anishinaabewi-gichigami is an oddity.
HA HA HA HA HA
Imagine this unpronouncable name on a trade contract!

Clueless colonizers. What an absurd idea to draw boundaries across the water... and yet they just get what they want! What a 'land of the free'!
Anyway, it's finally time for me to get out of here. Good riddance!

smarck Tribune.

...

This is just the beginning.
Turtle Island – Stolen Land
STOLEN LAND
SOVEREIGN INDIGENOUS LAND
#LANDBACK
Never stop learning...
...from the ones that have been silenced for too long.

Bismarck, 2018. Protests against the Dakota Access Pipeline
STAND WITH STANDING ROCK
NO DAPL
PROTECT WATER CLIMATE HEALTH
HONOR TREATY RIGH
Still, at the end of the day...
The hanitch-Huffman house now belongs to the Osage nation!
...we can never be 'one'.

May the bridges we burn light our way.

Author: Chandra-Milena Danielzik
Artist: Michel Esselbrügge

DOMINATION AND CONSTANT ACCUMULATION OF WEALTH ARE INTRINSIC TO CAPITALISM. SO ARE RACISM AND COLONIALISM. THEY ARE NOT SIMPLY ENTANGLED WITH CAPITALISM – RACISM IS ITSELF INTRINSIC TO THE HISTORIC AND PRESENT FORMATION OF CAPITALISM.

THAT'S ONE SMALL STEP FOR MAN, ONE GIANT LEAP FOR MANKIND.

Whitey on the moon

Racialised Inequality & Development as Destruction

LISTEN TO THE SONG, WHITEY! YOUR "MANKIND" HAS ALWAYS BEEN EXCLUSIVE! THE COLONISED, THE RACIALISED, THE LABOURING, THE CARE-WORKING MASSES HAVE NEVER BEEN PART OF YOUR MANKIND CLUB.

A rat done bit my sister Nell.
with Whitey on the moon
Her face and arms began to swell.
and Whitey's on the moon
I can't pay no doctor bill.
but Whitey's on the moon
Ten years from now I'll be payin' still.
while Whitey's on the moon
The man jus' upped my rent las' night.
'cause Whitey's on the moon
No hot water, no toilets, no lights.
but Whitey's on the moon
I wonder why he's uppi' me?
'cause Whitey's on the moon?
I was already payin' 'im fifty a week.
with Whitey on the moon
Taxes takin' my whole damn check,
Junkies makin' me a nervous wreck,
The price of food is goin' up,
An' as if all that shit wasn't enough
A rat done bit my sister Nell.
with Whitey on the moon
Her face an' arm began to swell.
but Whitey's on the moon
Was all that money I made las' year
for Whitey on the moon?
How come there ain't no money here?
Hm! Whitey's on the moon
Y'know I jus' 'bout had my fill
of Whitey on the moon
I think I'll sen' these doctor bills,
Airmail special
to Whitey on the moon

Gil Scott-Heron
Whitey on the moon, 1970

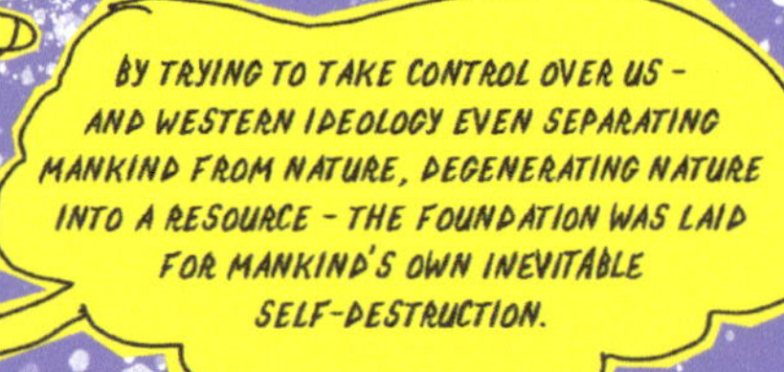

REMEMBER US? JUST DROPPING BY TO REMIND YOU OF THE COLONIAL PRESENT OF ENVIRONMENTAL DESTRUCTION AND CLIMATE CRISIS.

BY TRYING TO TAKE CONTROL OVER US – AND WESTERN IDEOLOGY EVEN SEPARATING MANKIND FROM NATURE, DEGENERATING NATURE INTO A RESOURCE – THE FOUNDATION WAS LAID FOR MANKIND'S OWN INEVITABLE SELF-DESTRUCTION.

HELLO? WHO IS THIS?

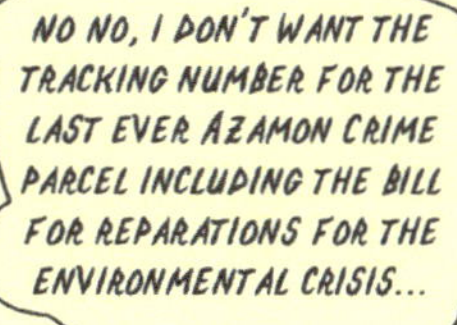

WHAT?! I'VE NEVER EVEN BEEN TO PAKISTAN AND ZAMBIA.

WHAT DO I HAVE TO DO WITH FLOODS AND DROUGHTS THERE? I THINK YOU GOT THE WRONG NUMBER...

WHAT DO YOU MEAN? ... IT'S THE RIGHT NUMBER BECAUSE EACH BILLIONAIRE' SPACE FLIGHT EMITS MORE CO2 THAN A NORMAL PERSON PRODUCES IN A LIFETIME. AND THE EMISSIONS FROM OUR LIFESTYLES ARE THOUSANDS OF TIMES THOSE OF AN AVERAGE PERSON, WHILE THE EMISSIONS FROM OUR INVESTMENTS ARE MORE THAN A MILLION HIGHER.

THEY TALK TO ME ABOUT PROGRESS, ABOUT "ACHIEVEMENTS", ... I AM TALKING ABOUT CULTURES TRAMPLED UNDERFOOT, ... LANDS CONFISCATED, ... EXTRAORDINARY POSSIBILITIES WIPED OUT. ... I AM TALKING ABOUT NATURAL ECONOMIES THAT HAVE BEEN DISRUPTED.

Aimé Césaire, Discourse on Colonialism, 1955

Tracking Trauma
German Genocides
at Home and Abroad
Author: Daniel Bendix / Artist: Hangula Werner
Go back a bit. Everything should be in the picture.

Fucking hell!!! I can't belive you dropped my phone!
FARM VREDEHEIM EST. 1910
I'm really sorry, baas. It's not broken, see ...
Keep your dirty fingers to yourself.
Ok,baas.

Pa, why do you let yourself be treated like that? As if we're still colonised.
You know what you do?! You actually support the colonisers, if you don't fight back. Just like our ancestors at the Battle of Waterberg in 1904.
When I grow up, I will make the tables turn.
My dear, it was much more complicated than that. Our ancestors were forced to provide military assistance to the Germans ....

Nama groups changed (or were forced to change) alliances in the context of German colonial aggression. The Witbooi (| Khowesin), for example, first fought against the OvaHerero, then made peace and fought side-by-side with the OvaHerero under German command against other colonised people, then fought against the OvaHerero for the Germans, until finally uniting with the OvaHerero and other Nama groups against the German occupiers.

If I were to confront Baas Jacobs, he'd fire me immediately.
But still ... This farm that we live on, the land was ours. It's still ours. Our people died for this land. You can't go on like that.
FARM VREDEHEIM EST. 1910

The Nama who had survived the genocide by the German colonisers of 1904-1908 were deprived of their land. The occupiers forced them into wage labour, to wear a badge around the neck for identification, and prevented them from keeping cattle and horses. Jobs with white settlers were often the only option to survive. At present, 70 percent of all private land in Namibia is still owned by a few thousand white and predominantly German-speaking farmers.
But you know! If I lose my job, we also lose the right to live here. Where else would we go?

NAMIBIA UNIVERSITY OF SCIENCE AND TECHNOLOGY
DEPARTMENT OF ARCHITECTURE
Pa... I am so happy to be home.
Did you meet any future husband on your trip to Cape Town?
Don't be ridiculous! It was a trip to study architecture, not men.

I wish I could also go to university after school.
Katrina, darling! You know we can't afford that ...
FARM VREDEHEIM EST. 1910
MAY
CLASS CANCELLED UNTILL FURTHER NOTICE!

Cape Town
Holocaust &
Genocide Centre
We went to the Holocaust Museum. It's a great building. We also went to see an exhibition there. You won't believe what I came across!
There was a picture of great-grandma in the exhibition! You never told me that she was Jewish and escaped the Holocaust.
Hi Frederick, good to see you. How are you?
I am well, Miss Frida.

All I know is that granddad met my grandma on a trip to Cape Town. He was there for some agricultural fair, I think. They bought this farm right after that.
Yessis, Frederick! Is it so hard to remember that I don't take lemon in my gin 'n tonic!!!
SS STUTTGART
Are you sure? I know she was German. She never spoke about her time before coming to Africa.
YAD VASHEM

You know, Frederick, I just found out that my great-grandmother had been forced to flee Nazi Germany in 1936. Her family lost everything, because they were Jewish. I wonder who stole their house, and who lives there now.
HAUS
FAMILIE
MOSER
My great-grandma, too! She had managed to flee to South Africa. But the Nazis imprisoned both her brothers for forced labour in a quarry.
Eventually, my great-granduncles were deported to a death camp called Sobibor. They must have either died on the way in the railroad carriages or been murdered in the extermination camp. The Germans murdered a quarter of a million Jews in that camp alone!

I wasn't aware that she was a genocide survivor, too, Miss Frida. All I knew was that my great-grandma worked for your great-grandparents as a servant. She had been the only one from our family to survive the genocide.
The Germans had forced my great-grandmother to build this railway between Lüderitz and Keetmanshoop. So many were worked to death! They used the railway to bring soldiers and war material to quash our resistance. Ever since, it has been key to stealing our resources.
My family had fought together with the Bethanian Nama Cornelius Frederiks. Then the Germans imprisoned them in th concentration camp on Shark Island. There, they forced great-grandma to scrape the flesh off the skulls of her own family!
people!
51

Weird how our histories are so similar and yet our present is so different.
Back to work. You piece of S!#."@T!
NAMIBIA UNIVERSITY OF SCIENCE AND TECHNOLOGY
DEPARTMENT OF ARCHITECTURE
Precolonial Namibian architecture...
Everybody loves to play memory, but *fokol* changes.
Of course we shouldn't forget. But the alternative is not necessarily remembering, but rather justice.

NAMA LANDLESS FARMERS ASSOCIATION VREDEHEIM
NAMIBIA LAND REFORM ELITE EMPOWERMENT PROGRAMME
Don't you drop my new phone!
Don't worry, Mr Commissioner, I've done this before.

Author: Chandra-Milena Danielzik
Artist: Michel Esselbrügge
Whose Cup of Tea
Migration, Colonialism & Plantation Capitalism
PEOPLE LIKE ME WHO CAME TO ENGLAND IN THE 1950S HAVE BEEN THERE FOR CENTURIES; SYMBOLICALLY, WE HAVE BEEN THERE FOR CENTURIES. I WAS COMING HOME.
I AM THE SUGAR AT THE BOTTOM OF THE ENGLISH CUP OF TEA.
I AM THE SWEET TOOTH, THE SUGAR PLANTATIONS THAT ROTTED GENERATIONS OF ENGLISH CHILDREN'S TEETH.
THERE ARE THOUSANDS OF OTHERS BESIDE ME THAT ARE, YOU KNOW, THE CUP OF TEA ITSELF. BECAUSE THEY DON'T GROW IT IN LANCASHIRE, YOU KNOW. NOT A SINGLE TEA PLANTATION EXISTS WITHIN THE UNITED KINGDOM.
THIS IS THE SYMBOLIZATION OF ENGLISH [EUROPEAN] IDENTITY – MEAN, WHAT DOES ANYBODY IN THE WORLD KNOW ABOUT AN ENGLISH PERSON EXCEPT THAT THEY CAN'T GET THROUGH THE DAY WITHOUT A CUP OF TEA? WHERE DOES IT COME FROM?
CEYLON, SRI LANKA, INDIA. THAT IS THE OUTSIDE HISTORY THAT IS INSIDE THE HISTORY OF THE ENGLISH [EUROPEAN]. THERE IS NO ENGLISH [EUROPEAN] HISTORY WITHOUT THAT OTHER HISTORY. Stuart Hall
STARBUCKS COFFEE
PROFITS OF CAPIT
THE ECONOMIC MOTIVATION THAT DRIVES POORER PEOPLE TO MIGRATE HAS BEEN PRODUCED AND CONTINUES TO BE REPRODUCED BY PRACTICES EMANATING FROM RICHER COUNTRIES AND THEIR OWN DEFICIENT UNDERSTANDINGS OF THEIR GLOBAL DOMINANCE.
EUROPE'S RELATIVELY HIGH STANDARD OF LIVING AND SOCIAL INFRASTRUCTURE HAVE NOT BEEN ESTABLISHED OR MAINTAINED SEPARATE FROM EITHER THE LABOUR AND WEALTH OF OTHERS, OR THE CREATION OF MISERY ELSEWHERE. Gurminder K. Bhambra
COLONIALISM AND IMMIGRATION ARE PART OF THE SAME CONTINUUM – WE ARE HERE BECAUSE YOU WERE THERE. THE SAME SYNDROME OBTAINS TODAY. EUROPE WANTS IMMIGRANT LABOUR BUT NOT THE IMMIGRANT, THE PROFIT FROM THE ONE, NOT THE COST OF THE OTHER …
IN FACT, GLOBALISATION AND IMMIGRATION ARE PART OF THE SAME CONTINUUM. WE ARE HERE BECAUSE YOU ARE THERE. Ambalavaner Sivanandan
Wish list
iPhone 20 Pro
Traditional matcha tea set
Ceylon Nuwara Sri Lankan Tea

space enthusiast
WE HAVE TO TAKE OUT LOANS TO PAY FOR FOOD AND WE STILL MISS MEALS REGULARLY
NOW:
2023
WE GIVE OUR BLOOD, SO THEY CAN LIVE COMFORTABLY
Rangasamy Puwaneshkanth
Subramaniam Sathyavani
THE PINCHING OF THE STOMACH IS MORALLY GOOD BECAUSE IT WILL INDUCE PEASANTS TO WORK ON PLANTATIONS
19th century British colonial analysts and data collectors on "Ceylon"
then:
COLONIAL CONTINUITY AT ITS BEST!
COLONIZER
Ceylon
Camellia sinensis
"Tea plant"
To Do List
Plantation Economy
Create a wage labour force, e.g.: by robbing people of their (communal) land, destroying subsistence farming, and increasing food prices so people are forced to work for you
Form capitalist and invent racialised relations between people: create racialised elites and divide the whole society into capitalist classes
Import Indian Tamils as cheap indentured labour, dehumanise them, and subject them to control by other colonised people
Manipulate caste relations and transform them to serve your plantation economy
Always make sure that race remains the dividing factor to avoid class consciousness or united anti-colonial struggle
Kill, starve, rape
Impoverishment
divided by race & class
Oppression and killing of Tamils & Muslims
subordinate integration into global capitalism
Export economy: Dependency on food imports
Starvation
Colonialism divided in order to rule what it integrated in order to exploit.
Ambalavaner Sivanandan
COLONIALISM
CAPITALISM
Wish list
Ceylon Nuwara Sri Lankan Tea
order now!
TAP!

Alienating the SDGs
A critique from outside
Author: Juan Telleria Zueco + Artist: Maite Mentxaka Tena
56

"The Sustainable Development Goals are a plan to save the world ... they are of...the people...

... by the people ... and for the people.

They will leave no one behind...

... and transform the world...

... for the benefit of all."

It's been a hard week. I'm so tired. But I guess, transforming the world is hard work. I'll be upstairs in my office.

installing e-alienated.exe (%90)
Accept
What?!?

e-alienated.exe installed

WO WOOO...
WWWOOOOOOOOO
WOOOOOOOOOOOaaaaaaaaaa
aaahhhaaaWoooooaaaa woo...

ahhaaa... woaahwoooo

ALiEN
UNIVERSITY
Campus

AUC

Hello everyone.
Please, open the book to page 365...
The earth:
a lost planet in the middle of nowhere, dominated by a very young species called humans.

They conquered and transformed the whole planet. Now they struggle with their own contradictions... to the extent that they could soon disappear!
For our practical session, we invited a specimen from the earth. Let us speak to them.

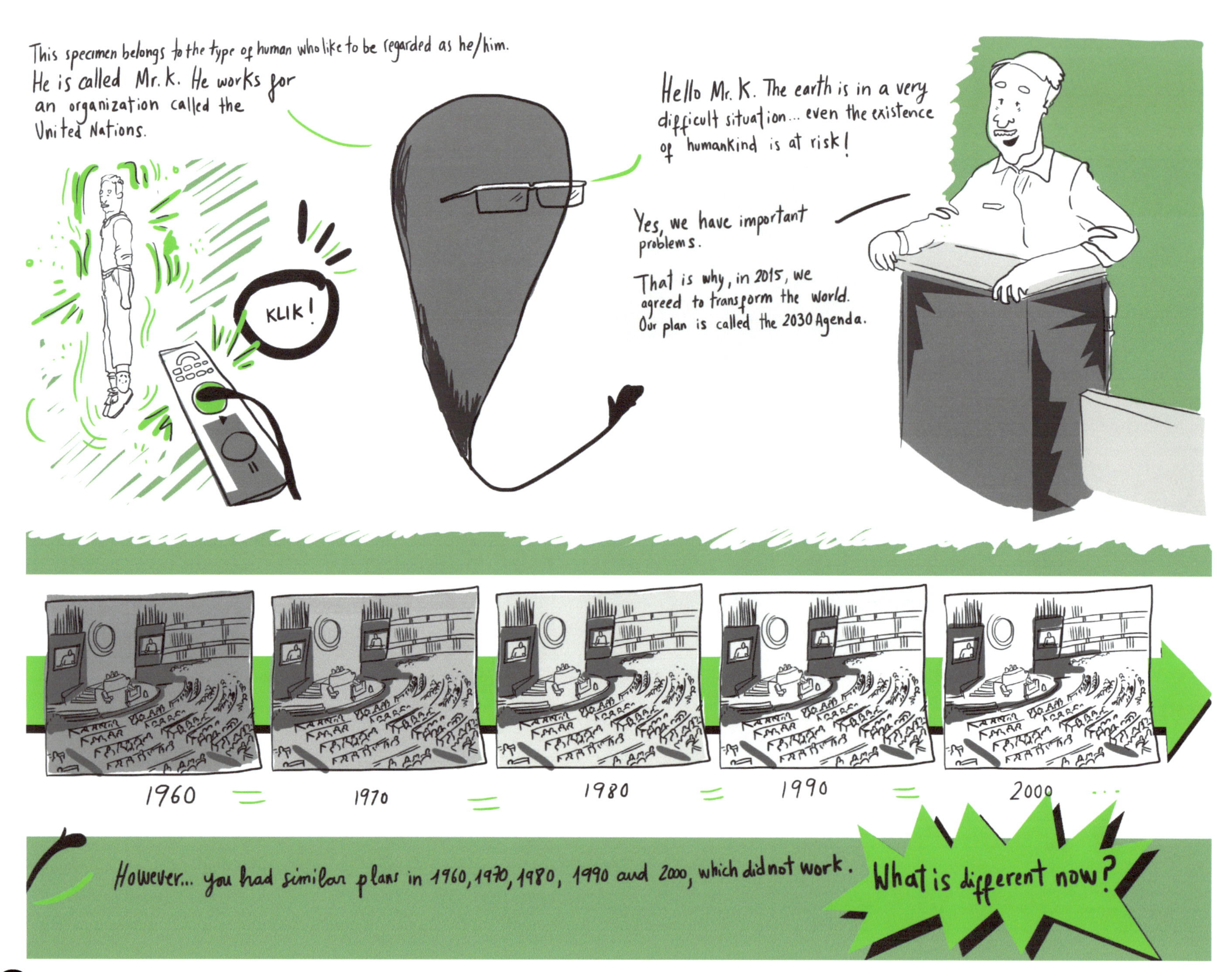
This specimen belongs to the type of human who like to be regarded as he/him.
He is called Mr. K. He works for an organization called the United Nations.
KLIK!
Hello Mr. K. The earth is in a very difficult situation... even the existence of humankind is at risk!
Yes, we have important problems.
That is why, in 2015, we agreed to transform the world. Our plan is called the 2030 Agenda.
1960
1970
1980
1990
2000 ...
However... you had similar plans in 1960, 1970, 1980, 1990 and 2000, which did not work.
What is different now?

For example, Mr. K.'s ancestors worked hard in making others understand their own universal truth.

When reason became the new God, they divided the world into different areas in order to teach others how to be rational, productive and efficient.

They even made rational and efficient USE OF NATURE, but it systematically benefited some at the expense of others.

You might not believe me: THEY DO NOT REALISE THAT ASSUMING A PARTICULAR PERSPECTIVE TO BE THE UNIVERSAL TRUTH SYSTEMATICALLY EXCLUDES OTHER PERSPECTIVES.

As Mr. K. explained, the 2030 Agenda reproduces the assumption that a set of universal principles exist, which will benefit all.

KLIK

What is the second important aspect, Mr. K.?

That the United Nations is the common house of humankind, which will help in creating a better world for all.

PAUSE

Sorry Mr. K! This is important, too:

Eight decades ago, a great war shook the world.

The groups with stronger weapons and bigger armies won, and created an organisation intended to stabilise the new world order:

As Mr. K. explained, the 2030 Agenda assumes that the organization created by the powerful ones is "the common house of humankind".

KLIK!

Your plan is very laudade, Mr. K. However, I don't think it will work.

Why not?

Because it reproduces the most bassic assumptions that took the earth to its current situation: the values of the powerful ones are assumed to be universal and beneficial for everyone.

What? Then how do you organise coexistence on the basis of such a relativistic perspective?

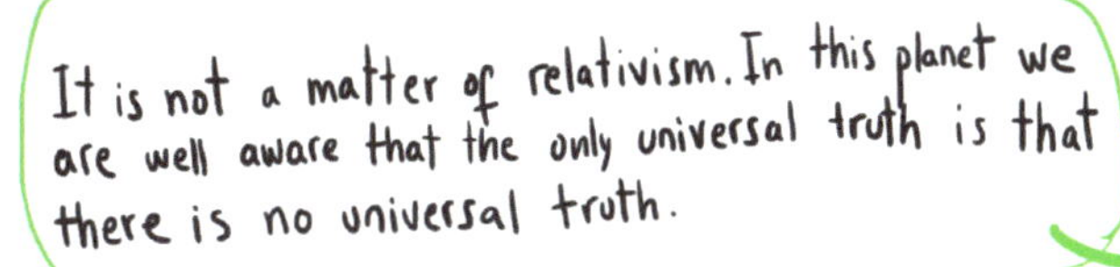

What?

We do not aim to materialise any universal truth. We just accept and negotiate our differences so we can coexist.

But that is not stable!

We are not afraid of contingency and change... we re-think, adapt and reconstruct our agreements and institutions every generation. It is not easy, but it works.

Oh... change? contingency? for every generation?... mmmmhhh...

hange

contingency

every generation

change

[A few days later]

- United Nations, New York -

SECRETARY GENERAL
The SDG promotional tour was good. In one of my ... mmmhh ... "visits", I had some important insights that could help in transforming the world. I wrote a report...
...Knock, knock, knock!

Mr. K.... You have been working with us for a long time... you know the UN is a serious institution with a long history... our common house, indeed... the SDGs are of the people, by the people and for the people... if we work together on the basis of our shared universal principles and values, we will transform the world and leave no one behind... In any case, I WILL READ YOUR REPORT CAREFULLY...

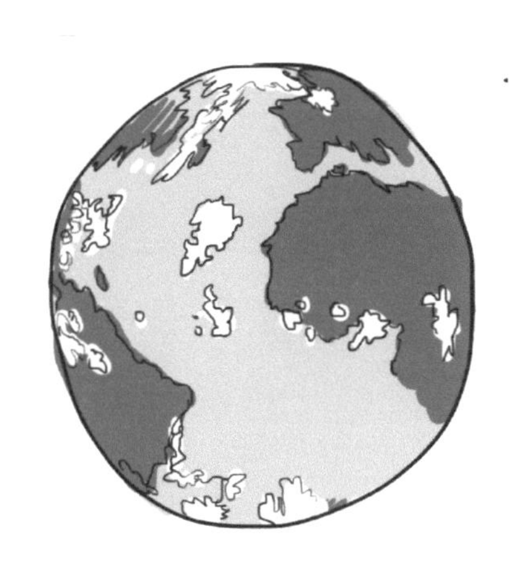

BLURRED IDENTITIES: CAN I BE THE COLONISER AND THE COLONISED?

Author: Lata Narayanaswamy
Artist: Michel Esselbrügge

SHE'S NOT INDIGENOUS!
BOOM!
BOOM!
BOOM!
Development NGO volunteer

THANK YOU!
(STRONG CANADIAN ACCENT)
Boom Boom!

BOOORN IN THE GLOBAL NORTH, I WAS, BOOORN IN THE GLOBAL NORTH!!!

10 YEARS LATER
HM ... MAYBE I SHOULD HAVE JUST LEFT THAT CLUB?
WHY DIDN'T I SPEAK OUT?

Under Development. Future Uncertain

AUTHORS: DANIEL BENDIX, LATA NARAYANASWAMY & ARAM ZIAI
ARTIST: ROSHNI VYAM

Mr. Leonard Consultorian
on his way to this year's
10th evaluation of a
development project for the
United Kingdom's Foreign,
Commonwealth & Development
Office

देहरादून हवाईअड्डा
DEHRADUN AIRPORT
Ah, it's been a while, I love Bangladesh!
This is India, Sir. Uttarakhand, to be precise, Sir.
Ah... of course, that's what I meant.

HOTEL GRAND
Good day. I am Laxmi. The NGO Aatmanirbhar, whose work on self-help you've been sent to evaluate, has assigned me as your interpreter.
Pleased to meet you. You can call me Leo!

This is the last of the 26 villages to be resettled for the dam. People engage in subsistence agriculture and gather food in the forest. Some grow sugarcane to sell on the market.
They are not rich, but they get by. They've been here for generations, it's their home. So obviously, they fear being evicted – and having to face the uncertainties of the city slums.

धरती माता
Fight For our land
Land Rights
Oh dear, I am sorry to hear that. That's awful.
I hope the community doesn't think we're going to get involved. It's not our job to interfere in local politics. We still need governments and the private sector as partners. .... Hopefully, she understands that I am really only here to do the interim evaluation and to check that the project will yield the expected outcomes. ... I only have three days ....

At least they are not dependent on the weather any more for making a living.
But will they be able to do that in the city? Dependent only on the market ...

beneficiaries ... met project targets ... log frame ... outcomes and impact ... theories of change ... gender checklist ... sustainability indicators ...

GYM
30 years treadmill,
30 years weights...
no wait: 30 minutes treadmill,
30 minutes weights!

It's all a hoax. Development is something people do to others less powerful than themselves.
Come on, the project at least helps them to cope with life in the modern world.
Yes, after they have been robbed of all they had before. I'm helping the development machine, not the people!

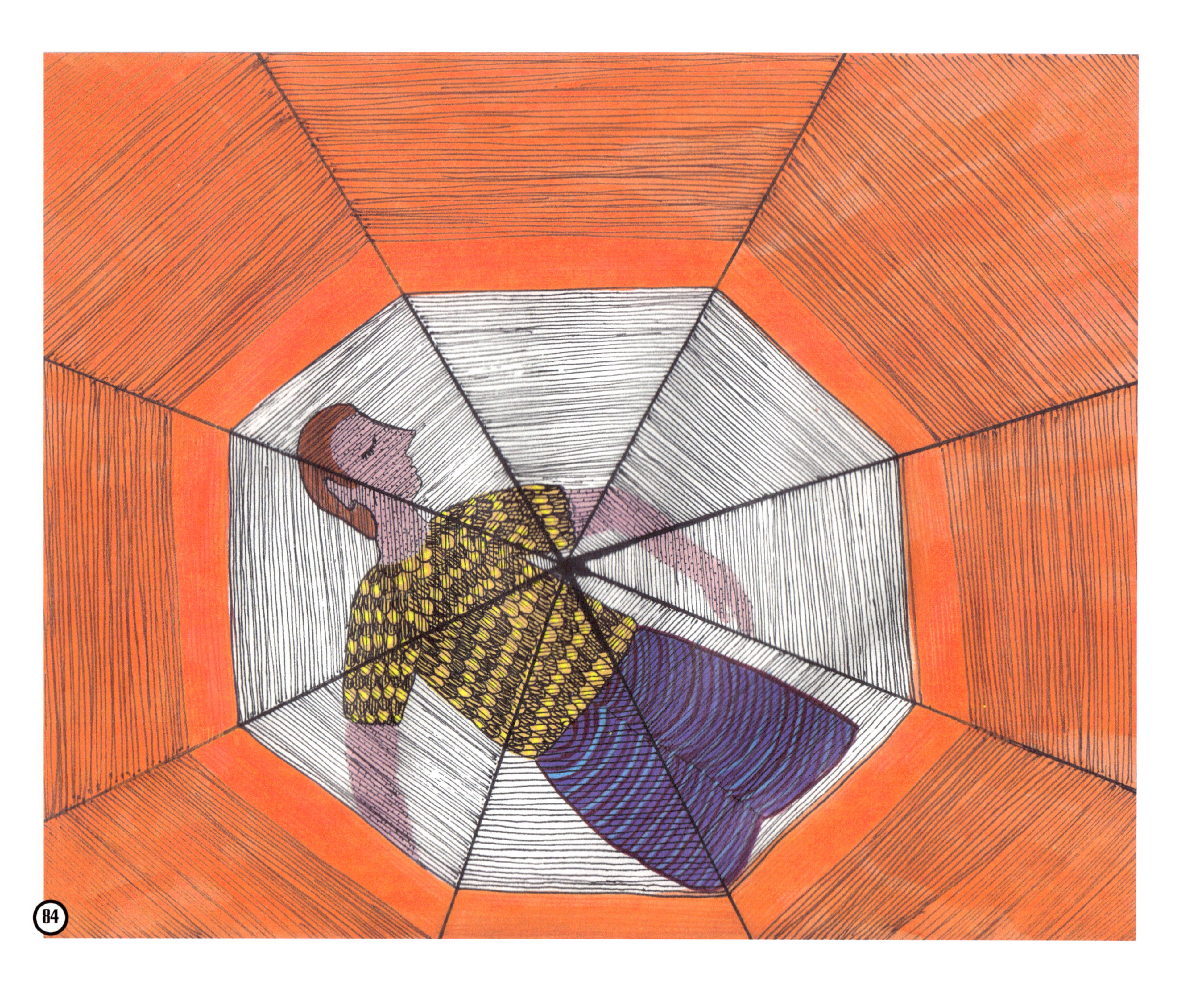

HSBC
Tate & Lyle

London Community-Supported Agriculture
It's been a while, Laxmi. How was your trip? With the new boats, it now only takes two weeks from India to the UK, no?
Honestly speaking, I never really believed in development.
Everybody is so grateful that you've accepted our invite to facilitate our transition to true sustainability.

You have the necessary knowledge, you just have to dig. It's buried under 500 years of tradition: that is, capitalism, colonialism and patriarchy.
Err... what? ... How do I do this? I can't ask my Grandma any more, can I?
So show us how this subsistence agriculture works. What are we supposed to do?

Anthropocentric blabla, if you ask me.
Yeah, it's still about themthemthem ... what about us??
Nobody said it would be easy.

Is this for digging?

Bibliography

Introduction and thanks

Cornwall, A. and Brock, K. (2006). The New Buzzwords, in P. Utting (Ed.), Reclaiming Development Agendas: Knowledge, Power and International Policy Making. Basingstoke, Palgrave Macmillan, 85–101.
Crenshaw, K. W. (1989). Demarginalizing the intersection of race and sex: A black feminist critique of antidiscrimination doctrine, feminist theory, and antiracist politics. University of Chicago Legal Forum, (1), 139–167.
Hayter, T. (1971). Aid as Imperialism. Harmondsworth, Penguin.
Hickel, J. (2017). The Divide. A brief guide to global inequality and its solutions. London, Penguin.
Hill Collins, P. (1990). Black Feminist Thought: Knowledge, Consciousness and the Politics of Empowerment. Boston, Unwin Hyman.
hooks, b. (1981). Ain't I a Woman? Black Women and Feminism. Boston, South End Press.
Mohanty, C. T. (1988). Under Western Eyes: Feminist Scholarship and Colonial Discourses. Feminist Review, (30), 61–88. https://doi.org/10.2307/1395054
Nkrumah, K. (1965). Neo-Colonialism, the Last Stage of Imperialism. London, Thomas Nelson & Sons.
Quijano, A. (2000). Coloniality of Power, Eurocentrism, and Latin America. Nepantla: Views from the South, 1 (3), 533–580.
Rothberg, M. (2009). Multidirectional Memory: Remembering the Holocaust in the Age of Decolonization. Oxford, Oxford University Press.
SAPRIN (Structural Adjustment Participatory Review Initiative) (2004). Structural Adjustment. The Policy Roots of Economic Crisis and Poverty. London, Zed Books.
Spivak, G. C. (1988). Can the Subaltern Speak? in C. Nelson and L. Grossberg (Eds.), Marxism and the Interpretation of Culture. London, Basingstoke, 271–313.
UN General Assembly (1960). Declaration on the granting of independence to colonial countries and peoples. Resolution 1514 (XV).

Never Conquered. On the Destruction and Creation of Knowledge

CHERÁN K'ERI. https://cheran.gob.mx
Civic Council of Popular and Indigenous Organizations of Honduras (COPINH). https://copinh.org/en/
Comandanta Esther in the Congress of the Union (2001). https://schoolsforchiapas.org/wp-content/uploads/2014/03/Comandanta-Esther-in-the-Congress-of-the-Union.pdf
First Declaration of the Lacandon Jungle (1993): https://radiozapatista.org/?p=20280&lang=en
Front Line Defenders. https://www.frontlinedefenders.org/en/location/kenya
Gruppe Asamblea Berlin (2024). Dass ich den Kalender meiner Vorfahren nutzen kann, bewegt mich sehr. Ein antikolonialer Besuch der Dresdner Landesbibliothek. ILA Bonn, 475, https://www.ila-web.de/ausgaben/475/dass-ich-den-kalender-meiner-vorfahren-nutzen-kann-bewegt-mich-sehr
Lugones, M. (2007). Heterosexualism and the Colonial/Modern Gender System. Hypatia, 22 (1), 186–219. doi:10.1111/j.1527-2001.2007.
Lugones, M. (2010). Toward a Decolonial Feminism. Hypatia, 25 (4), 742–759. doi:10.1111/j.1527-2001.2010.01137.x.
Nosotr@s (2021). Primera Parte: UNA DECLARACIÓN POR LA VIDA. http://enlacezapatista.ezln.org.mx/2021/01/01/primera-parte-una-declaracion-por-la-vida/
Producciones y Milagros Agrupación Feminista. https://www.instagram.com/produccionesymilagros/
Q'eqchi' El Estor Resiste. https://www.facebook.com/EstorResiste/
Reed, N.A. (2001). The Caste War of Yucatan. Stanford, Stanford University Press.
Solís Mecalco, R. de J. (2020). Dekolonisierung der Maya-Sexualitäten im Südosten Mexikos. PERIPHERIE, 40 (157/158), 81–101. https://doi.org/10.3224/peripherie.v40i1-2.05
Research: Tren "Maya" Made in Germany (2022). https://www.ya-basta-netz.org/research-tren-maya-made-in-germany/
RIP.MX.HAGAMOS GUERILLA CULTURAL (2021). https://rip.mx/one-hundred-days-of-otomi-hope/
Rosado Rosado, G. and Santana Rivas, L. (2008). María Uicab: reina, sacerdotisa y jefa militar de los mayas rebeldes de Yucatán (1863-1875). Mesoamérica, 29 (50), 112–139.
Rosado, Rosado G. and Chablé Mendoza, C. (2020). En busca de María Uicab: Reina y Santa patrona de los mayas rebeldes. Mexico City, Ediciones el nido del fénix.
Samir Vive. El Altar. https://www.ahuehuete.org/samirvive
SupGaleano (2021). 421st Squadron. https://enlacezapatista.ezln.org.mx/2021/04/20/421st-squadron/
Tzul Tzul, G. (2016). Sistemas de gobierno comunal indígena. Mujeres y tramas de parentesco en Chuimeq'ena'. Editorial. Guatemala City, SOCEE/Maya'Wuj.
Wikipedia (2024). Cherán. https://en.wikipedia.org/wiki/Cher%C3%A1n

Savages Setting Sail. Gender, Sexual Violence & Colonialism

Clarke, J. H. (1992). Christopher Columbus and the Afrikan Holocaust. Slavery and the Rise of European Capitalism. New York, A & B Books.
Danielzik, C.-M. and Bendix, D. (2011). 'Entdecken / Entdeckung / Entdecker_in / Entdeckungsreise'. In S. Arndt and N. Ofuatey-Alazard (Eds.), Wie Rassismus aus Wörtern spricht. (K)Erben des Kolonialismus im Wissensarchiv deutscher Sprache. Ein kritisches Nachschlagewerk. Münster, Unrast, 264–268.
Deer, S. (2015). The Beginning and End of Rape. Confronting Sexual Violence in Native America. Minneapolis, Minnesota University Press. https://www.upress.umn.edu/book-division/books/the-beginning-and-end-of-rape.
de Las Casas, B. (1552). A Brief Account of the Destruction of the Indies. 1552. https://faculty.chass.ncsu.edu/slatta/hi216/documents/dlascasas.htm.

Ensminger, J. (2022). From Hunters to Hell Hounds: The Dogs of Columbus and Transformations of the Human-Canine Relationship in the Early Spanish Caribbean. Colonial Latin American Review, 31 (3), 354–380. https://doi.org/10.1080/10609164.2022.2104035.
McClintock, A. (1995). Imperial Leather: Race, Gender, and Sexuality in the Colonial Contest. London, Routledge.
Native Americans in Philanthropy and Candid. (2023). Investing in Native Communities. https://nativephilanthropy.candid.org/native-101.
Ritter, D. (2020). The Rejection of Terra Nullius in Mabo: A Critical Analysis. The Sydney Law Review, 18 (1), 5–33. https://doi.org/10.3316/ielapa.970100936.
Shah, S. (2021). 'The Doctrine of Discovery and Terra Nullius'. The Indigenous Foundation (Blog). 25 October 2021. https://www.theindigenousfoundation.org/articles/the-doctrine-of-discovery-and-terra-nullius.
Sjursen, M. D. (2018). American History for Truthdiggers: Whose Revolution? (1775-1783). Truthdig, 7 April 2018. https://www.truthdig.com/articles/american-history-for-truthdiggers-whose-revolution-1775-1783/.
Smith, A. (2003). Not an Indian Tradition: The Sexual Colonization of Native Peoples. Hypatia, 18 (2), 70–85.
Stark, C. and Hudon, E. (2020). Colonization, Homelessness, and the Prostitution and Sex Trafficking of Native Women. Harrisburg: National Resource Center on Domestic Violence. https://www.niwrc.org/sites/default/files/images/resource/colonization-homelessness-nativewomen.pdf.
United Nations Economic and Social Council. (2012). 'Doctrine of Discovery, Used for Centuries to Justify Seizure of Indigenous Land, Subjugate Peoples, Must Be Repudiated'. https://press.un.org/en/2012/hr5088.doc.htm.

It All Runs in the Family

Appropriation (2021). Nothing for Ourselves. https://www.youtube.com/watch?v=XNxGasluC_k
Benally, Klee (2023). No Spiritual Surrender: Indigenous Anarchy in Defense of the Sacred. Detritus Books.
Bismarck Tribune. (1884). A Pet Alligator, 22.2.1884 http://dakotadeathtrip.com/Alligator_1423870859.html
Bismarck Tribune (1891). "Louis Hanitch Talks of West Superior as a Wheat Shipping Point", 20.11. 1891, https://chroniclingamerica.loc.gov/lccn/sn85042588/1891-11-20/ed-1/seq-5/#words=%22that+is+a+lot+of%22
Gilio-Whitaker, D. (2018). Settler Fragility: Why Settler Privilege Is So Hard to Talk About, https://www.beaconbroadside.com/broadside/2018/11/settler-fragility-why-settler-privilege-is-so-hard-to-talk-about.html
Hele, K. S. (Ed.) (2008). Lines drawn upon the water: First Nations and the Great Lakes borders and borderlands. Waterloo, Wilfrid Laurier University Press.
Herold, K. (2016). Terra Nullius and the History of Broken Treaties at Standing Rock. https://intercontinentalcry.org/terra-nullius-history-broken-treaties-standing-rock/
Karuka, M. (2019). Empire's Tracks: Indigenous Nations, Chinese Works, and the Transcontinental Railroad. Oakland, University of California Press.
Lahti, J. (Ed.) (2021). German and United States Colonialism in a Connected World: Entangled Empires. Berlin, Springer Nature.
Nelson, R.L. (2015). A German on the Prairies: Max Sering and settler colonialism in Canada, Settler Colonial Studies, 5 (1), 1–19, http://dx.doi.org/10.1080/2201473X.2014.899551
Supreme Court of United States.(1919). Minnesota v. Wisconsin, 252 U.S. 273 https://www.courtlistener.com/opinion/99543/minnesota-v-wisconsin/?q=Minnesota%20Wisconsin%201919
The Indigenous Digital Archive. IDA Treaties Explorer https://digitreaties.org/

Whitey on the Moon. Racialised Inequality & Development as Destruction

Bhattacharyya, G. (2018). Rethinking Racial Capitalism: Questions of Reproduction and Survival. London, Rowman & Littlefield International.
Césaire, A. (2001). Discourse on Colonialism. New York, Monthly Review Press.
Chakrabarty, D. (2000). Provincializing Europe. Princeton, Princeton University Press.
Duncan, J. S. (2016). In the Shadows of the Tropics: Climate, Race and Biopower in Nineteenth Century Ceylon. London, Routledge.
Federici, S. (2004). Caliban and the Witch: Women, the Body and Primitive Accumulation. New York, Autonomedia.
Hall, S. (1996). The West and the Rest: Discourse and Power. In S. Hall, D. Held, D. Hubert, and K. Thompson (Eds.), Modernity: An Introduction to Modern Societies. New York, Wiley-Blackwell, 185–227.
Kößler, R. (1998). Entwicklung. Münster, Westfälisches Dampfboot.
Leys, C. (2009). The Rise and Fall of Development Theory. Bloomington, Indiana University Press.
Matthews, S. (2004). Post-Development Theory and the Question of Alternatives: A View from Africa. Third World Quarterly, 25 (2), 373–84.
McClintock, A. (1992). The Angel of Progress: Pitfalls of the Term "Post-Colonialism". Social Text, (31/32), 84–98. https://doi.org/10.2307/466219.
McEwan, C. (2009). Postcolonialism and Development. New York, Routledge.
Melber, H. (1992). Der Weißheit letzter Schluß. Rassismus und kolonialer Blick. Frankfurt/Main, Brandes & Apsel.
Oxfam International. (2022). 'A Billionaire Emits a Million Times More Greenhouse Gases than the Average Person'. Oxfam International. https://www.oxfam.org/en/press-releases/billionaire-emits-million-times-more-greenhouse-gases-average-person.
Robinson, C. J. (2021). Black Marxism. The Making of the Black Radical Tradition. Chapel Hill, University of North Carolina Press.

Scott-Heron, G. (1970). Whitey on the Moon. Small Talk at 125th and Lenox.
Shiva, V. (2016). Staying Alive: Women, Ecology, and Development. Berkeley, North Atlantic Books.

Tracking Trauma: German Genocides at Home and Abroad

Cape Town Holocaust and Genocide Centre (2020). Seeking Refuge: German Jewish immigration to the Cape in the 1930s. Cape Town, 15/12/2020-22/01/2021.
Dookoom (2014). Larney Jou Poes. On A Gangster Called Big Times. https://dookoom.bandcamp.com/album/a-gangster-called-big-times
Erichsen, C. W. (2004). Zwangsarbeit im Konzentrationslager auf der Haifischinsel. In J. Zimmerer and J. Zeller (Eds.), Völkermord in Deutsch-Südwestafrika. Der Kolonialkrieg in Namibia und seine Folgen. Berlin, Ch. Links Verlag, 80–85.
Fanon, F. (1963). The Wretched of the Earth. New York, Grover Press.
Hillebrecht, W. (2004). Die Nama und der Krieg im Süden. In J. Zimmerer and J. Zeller (Eds.), Völkermord in Deutsch-Südwestafrika. Der Kolonialkrieg in Namibia und seine Folgen. Berlin, Ch. Links Verlag, 121–133.
Hillebrecht, W. (2015). Hendrik Witbooi and Samuel Maharero: The Ambiguity of Heroes. In J. Silvester (Ed.), Re-Viewing Resistance in Namibia History. Windhoek, University of Namibia Press, 38–54.
Luipert, S. (2021). The Herero-Namaqua genocide told by a survivor's descendent. https://www.youtube.com/watch?v=lfcyL60YIPU
Melber, H. (2022). Germany and reparations: The reconciliation agreement with Namibia. The Round Table, 111 (4), 475–488. https://doi.org/10.1080/00358533.2022.2105540
Nama Traditional Leaders' Association (NTLA). https://www.facebook.com/people/Nama-Traditional-Leaders-Association/100069848481063/
No Amnesty on Genocide! http://genocide-namibia.net/
Ovaherero Traditional Authorities (OTA). https://www.facebook.com/groups/712992288767165/
Rothberg, M. (2009). Multidirectional Memory: Remembering the Holocaust in the Age of Decolonization. Oxford, Oxford University Press.
Timm, U. (2000). Morenga. München, Deutscher Taschenbuch Verlag.
Yerushalmi, Y. H. (1996). Zakhor: Jewish History and Jewish Memory. Seattle, University of Washington Press.

Whose Cup of Tea. Migration, Colonialism & Plantation Capitalism

Balachandran, P. K. (2020). 'How Colonialism Crippled Sri Lankan Peasant Agriculture'. The Citizen - Independent Journalism, 24 November 2020. https://www.thecitizen.in/index.php/en/NewsDetail/index/6/19654/How-Colonialism-Crippled-Sri-Lankan-Peasant-Agriculture-.
Bandarage, A. (2020). Colonialism in Sri Lanka: Export Agriculture in Ceylon, 1833–1886. In A. Bandarage (Ed.), Colonialism in Sri Lanka. Boston, Vimukti, 65–86. https://doi.org/10.1515/9783110838640.
Bhambra, G. K. (2015). 'Europe Won't Resolve the "Migrant Crisis" until it Faces its Own Past'. The Conversation, 1 September 2015. http://theconversation.com/europe-wont-resolve-the-migrant-crisis-until-it-faces-its-own-past-46555.
Bhambra, G. K (2022). 'Colonial Taxes Built Britain. That Must Be Taught in Lessons on Empire'. Https://Dhakacourier.Com.Bd/, 8 April 2022. https://dhakacourier.com.bd/news/Essays/Colonial-taxes-built-Britain-That-must-be-taught-in-lessons-on-Empire/4764.
Casas-Cortes, M., Cobarrubias, S., and Pickles, J. (2015). Riding Routes and Itinerant Borders: Autonomy of Migration and Border Externalization. Antipode, 47 (4), 894–914. https://doi.org/10.1111/anti.12148.
Ha, K. N. (2003). Die kolonialen Muster deutscher Arbeitsmigrationspolitik. In H. Steyerl and E. G. Rodríguez (Eds.) Spricht die Subalterne Deutsch? Migration und postkoloniale Kritik. Münster, Unrast, 56–107.
Hall, S. (2004). Cultural Studies - Ein politisches Theorieprojekt: Ausgewählte Schriften 3. Hamburg, Argument.
Internationalist Communist Union. (2006). Sri Lanka - the Poisoned Legacy of British Colonial Rule https://www.union-communiste.org/en/internationalist-communist-forum/77-sri-lanka-the-poisoned-legacy-of-british-colonial-rule.
Kanapathipillai, V. (2009). Citizenship and Statelessness in Sri Lanka: The Case of the Tamil Estate Workers. London, Anthem Press.
Lakhiani, K. (2021). The Exploitation of Tea Plantation Workers in Sri Lanka, 22 August 2021. https://borgenproject.org/tea-plantation-workers-in-sri-lanka/.
Rappaport, E. (2017). A Thirst for Empire. How Tea Shaped the Modern World. Princeton, Princeton University Press.
Ravindran, J. (2023). 'We Give Our Blood so They Live Comfortably': Sri Lanka's Tea Pickers Say They Go Hungry and Live in Squalor'. The Guardian, 23 May 2023. https://www.theguardian.com/global-development/2023/may/23/we-give-our-blood-so-they-live-comfortably-sri-lankas-tea-pickers-say-they-go-hungry-and-live-in-squalor.
Sarvan, C. (2019). Indian Plantation ('Coolie') Experiences Overseas. Kunapipi, 22 (2), 8–27 https://ro.uow.edu.au/kunapipi/vol22/iss2/5.

Alienating the SDGs. A Critique from Outside

Derrida, J. (1981). Positions. Chicago, University of Chicago Press.
Escobar, A. (1995). Encountering Development. The Making and Unmaking of the Third World. Princeton, Princeton University Press.
Howarth, D. (2000). Discourse. Buckingham, Open University Press.
Laclau, E. (1996). Emancipation(s). London, Verso.
Mouffe, C. (1993). The Return of the Political. London, Verso.
Rist, G. (1996). The History of Development: From Western Origins to Global Faith. London, Zed Books.
Ziai, A. (2016). Development Discourse and Global History: From Colonialism to the Sustainable Development Goals. London, Routledge.

Blurred Identities. Can I be the Colonizer and the Colonized?

Bandyopadhyay, R. and Patil, V. (2017). 'The white woman's burden' – the racialized, gendered politics of volunteer tourism. Tourism Geographies, 19 (4), 644–657, https://doi.org/10.1080/14616688.2017.1298150

Blum, A. and Schäfer, D. (2017). Volunteer work as a neocolonial practice – racism in transnational education. Transnational Social Review, 8 (2), 155–169. https://doi.org/10.1080/21931674.2017.1401427

Boatcă, M. (2021). Thinking Europe otherwise: lessons from the Caribbean. Current Sociology, 69 (3), 389–414. https://doi.rg/10.1177/0011392120931139

Burlyuk, O. and Rahbari, L. (Eds.) (2023). Migrant Academics' Narratives of Precarity and Resilience in Europe. Cambridge, OpenBook Publishers, www.openbookpublishers.com/books/10.11647/obp.0331.

Chakrabarty, D. (2000). Provincializing Europe: Postcolonial Thought and Historical Difference. New Jersey, Princeton University Press.

Icaza, R. and Sheik, Z.B. (2023). When we couldn't breathe. (Our) stories from the margins. Globalizations, 20 (2), 201–207. https://doi.org/10.1080/14747731.2023.2184169

Le Bourdon, M. (2022). Confronting the discomfort: a critical analysis of privilege and positionality in development. International Journal of Qualitative Methods, (21), 1–8. https://doi.org/10.1177/16094069221081362

Nagar, R. (2014). Muddying the Waters. Coauthoring Feminisms Across Scholarship and Activism. Springfield, University of Illinois Press.

Narayanaswamy, L. (2024). 'Race, racialisation, and coloniality in the humanitarian aid sector'. In S. Roth, B. Purkayastha, and T. Denskus (Eds), Handbook on Humanitarianism and Inequality. London, Edward Elgar, 210–221.

Narayanaswamy, L. and Schöneberg, J. (2020). Introduction to the special focus. Acta Academica, 52 (1), 1–9.

Telles, E. and Flores, R. (2013). Not Just Color: Whiteness, Nation, and Status in Latin America'. Hispanic American Historical Review, 93 (3), 411–449.

Two Convivial Thinkers (2023). (Un)Doing performative decolonisation in the global development 'imaginaries' of academia. Global Discourse, 14(2-3), 355–379, 1–25, https://doi.org/10.1332/20437897Y2023D000000010

Ulu , Ö.M. and Bilgen, A. (2022). Some scholars are more equal than others. Social Psychological Review, 24 (2), 5–18.

Vrasti, W. (2013). Volunteer Tourism in the Global South: Giving Back in Neoliberal Times. London, Routledge.

Under Development. Future Uncertain

Burkhart, C., Schmelzer, M., and Treu, N. (Eds.) (2020). Degrowth in Movement(s). Exploring Pathways for Transformation. Winchester, John Hunt.

Césaire, A. (1953). Discours sur le colonialisme. Paris, Présence Africaine.

de Wet, C. (2006). Development-Induced Displacement: Problems, Policies, and People. Oxford, Berghahn.

Ferguson, J. (1994). The Anti-Politics Machine. 'Development', Depoliticization and Bureaucratic Power in Lesotho. Minneapolis, University of Minnesota Press.

Frank, L. (1997). 'The Development Game', in: M. Rahnema with V. Bawtree (Eds.), The Post-Development Reader. London, Zed Books, 263–273.

Kothari, A., Salleh, A., Escobar, A., Demaria, F., and Acosta, A. (Eds.). (2019). Pluriverse: A Post-Development Dictionary. Delhi, Tulika Books.

Li, T. M. (2007). The Will to Improve. Governmentality, Development, and the Practice of Politics. Durham, Duke University Press.

Rich, B. (1994). Mortgaging the Earth. The World Bank, Environmental Impoverishment and the Crisis of Development. Washington, Island Press.

Roy, A. (1999). The Greater Common Good. World Watch, 14 (1), 33–36. http://www.narmada.org/gcg/gcg.html

Sachs, W. (Ed.) (2010). The Development Dictionary. A Guide to Knowledge as Power. London, Zed Books.

Vyam, D., Vyam, S., Natarajan, S., and Anand, S. (2011). Bhimayana. Delhi, Navayana Publishing.

Ziai, A., Müller, F., and Bendix, D. (2019). Postdevelopment Alternatives in the North In E. Klein, E. and C. E. Morreo (Eds.), Postdevelopment in Practice: Alternatives, Economies, Ontologies. London, Routledge, 133–148.

www.ingramcontent.com/pod-product-compliance
Lightning Source LLC
LaVergne TN
LVHW070732120225
803493LV00036BA/315